INTELLIGENCE AND COUNTERINTELLIGENCE STUDIES

INTELLIGENCE IN CONTEMPORARY MEDIA

INTELLIGENCE AND COUNTERINTELLIGENCE STUDIES

Additional books in this series can be found on Nova's website under the Series tab.

Additional E-books in this series can be found on Nova's website under the E-book tab.

MEDIA AND COMMUNICATIONS - TECHNOLOGIES, POLICIES AND CHALLENGES

Additional books in this series can be found on Nova's website under the Series tab.

Additional E-books in this series can be found on Nova's website under the E-book tab.

INTELLIGENCE AND COUNTERINTELLIGENCE STUDIES

INTELLIGENCE IN CONTEMPORARY MEDIA

FRANCES M. WALSCH
EDITOR

Nova Science Publishers, Inc.
New York

For permission to use material from this book please contact us:
Telephone 631-231-7269; Fax 631-231-8175
Web Site: http://www.novapublishers.com

LIBRARY OF CONGRESS CATALOGING-IN-PUBLICATION DATA

Intelligence in contemporary media / editors [sic], Frances M. Walsch.
p. cm.
Originally published in the journal, Studies in intelligence: special review supplement, vol. 53, no. 2, Summer 2009.
Includes index.
ISBN 978-1-61122-651-5 (softcover)
1. Spy films--United States--History and criticism. 2. Intelligence service in motion pictures. 3. American fiction--20th century--History and criticism. 4. Intelligence service in literature. 5. Spy television programs--United States--History and criticism. I. Walsch, Frances M. II. Studies in intelligence.
PN1995.9.S68I58 2011
791.43'6556--dc22

2010042615

Published by Nova Science Publishers, Inc. † New York

Contents

PREFACE

The Center for the Study of Intelligence (CSI) was founded in 1974 in response to Director of Central Intelligence James Schlesinger's desire to create within the CIA an organization that could "think through the functions of intelligence and bring the best intellects available to bear on intelligence problems." The center, comprising professional historians and experienced practitioners, attempts to document lessons learned from past activities, to explore the needs and expectations of intelligence consumers, and to stimulate serious debate about current and future intelligence challenges. This book examines the numerous books and monographs addressing historical, operational, doctrinal and theoretical aspects of the intelligence profession and its relation to modern media.

One of the least appreciated facts about the intelligence profession is that it exists in, and is influenced by, a very complex environment—one that includes everything from its relationships with policymakers, legislatures, military services, foreign partners, and last but not least, to its interaction with the public. How intelligence relates to all of these arenas—and how it is regarded within them—ultimately affects everything from intelligence performance to funding to recruitment of personnel.

The public is a particularly important part of this environment. But unlike military services, intelligence organizations do not have recruitment centers in every mid-sized town; nor do most families have some member who has served in intelligence. Hence, what most in the public think about intelligence depends to a large extent on what they see in cinematic, documentary, and novelistic sources like those reviewed in this issue. This is particularly the case in the United States, but I suspect it is true by varying degree in all of the countries our reviewers represent or have spent time in.

This is an edited, reformatted and augmented version of Studies in Intelligence in Contemporary Media View of the Intelligence officers introduction by John McLaughlin, dated Summer 2009.

In: Intelligence in Contemporary Media
Editors: Frances M. Walsch
ISBN: 978-1-61122-651-5

Chapter 1

VIEW OF INTELLIGENCE OFFICERS

INTRODUCTION

John McLaughlin

One of the least appreciated facts about the intelligence profession is that it exists in, and is influenced by, a very complex environment—one that includes everything from its relationships with policymakers, legislatures, military services, foreign partners, and last but not least, to its interaction with the public. How intelligence relates to all of these arenas—and how it is regarded within them—ultimately affects everything from intelligence performance to funding to recruitment of personnel.

The public is a particularly important part of this environment. But unlike military services, intelligence organizations do not have recruitment centers in every mid-sized town; nor do most families have some member who has served in intelligence. Hence, what most in the public think about intelligence depends to a large extent on what they see in cinematic, documentary, and novelistic sources like those reviewed in this issue. This is particularly the case in the United States, but I suspect it is true by varying degree in all of the countries our reviewers represent or have spent time in.

As the reviewers make clear, what the public sees and reads is with rare exception fantasy mixed with a few kernels of truth. This is particularly true when it comes to American authors and directors. We have not yet produced an espionage novelist with the maturity and perfect pitch so frequently found in the work of British masters such as John le Carré—although American writers such as Charles McCarry and David Ignatius are edging into that circle.

Why is this not a better developed tradition in the United States? Part of the answer has to be that espionage is still a very new experience for us and that we are still a very young country. By contrast, countries such as Britain, France, and Russia have practiced spycraft in an organized way for centuries—and the Chinese strategist Sun Tzu was writing in sophisticated ways about espionage in the 6th century BC.

We did not organize intelligence formally on the national level until 1947. So it is hardly surprising that authors and directors in the United States are still coming to terms with this most arcane of instruments in the national security tool kit. And they are doing so, after all, during the period—the years since the 9/11 attacks—when the US polity itself has been reflecting deeply but inconclusively on what it wants from intelligence, what it doesn't want, and how to organize the effort most effectively.

As important as detail is to all art, what often lingers is a dominant impression—what you remember days, weeks, or years after seeing a film or reading a novel. And in that respect, what I take from these reviews is modestly reassuring. When you look, for example, at all of the recent films about espionage and subtract the pyrotechnics and action figure antics that strain credulity, what lingers are several broad impressions: intelligence officers are often courageous, prepared to stand on principle, forced frequently to deal with stressful or ambiguous circumstances, and quite willing to take risks. This is heartening.

Equally so is something that a real scholar of the genre, Ohio University Professor Emeritus of History Alan Booth, pointed out to his students: a great deal of complexity is embedded in the literature of espionage. As his course syllabus makes clear, various films and novels capture elements of heroism, ambiguity, illusion, and farce. For those of us who recoil against the oversimplification of the profession in the sound bite world of cable TV and modern journalism, this too is encouraging.

So enjoy the Intelligence in Contemporary Media section of this issue of *Studies in Intelligence* and take heart from the thought that if life truly does imitate art, that would not be an entirely bad thing in the case of espionage.

John McLaughlin is a former Deputy Director of Central Intelligence and former acting Director of Central Intelligence.

In: Intelligence in Contemporary Media
Editors: Frances M. Walsch
ISBN: 978-1-61122-651-5

Chapter 2

Intelligence in Fictional Literature

Stephen Maturin: The Ideal Intelligence Officer for Our Times

Nicholas Dujmovic

Intelligence professionals of the 21st century, particularly those engaged in the counterterrorism mission, could use an inspirational icon from the world of literature to replace the outdated figure of James Bond. It may come as a surprise that the best fictional figure to represent the kind of intelligence officer we need these days is a Catalan-Irish physician and naturalist serving the British Crown in the Royal Navy two centuries ago during the Napoleonic era.

Stephen Maturin is one of the two main characters in Patrick O'Brian's epic, 21-volume series set in the years 1800 through 1815 during the global struggle for power between Britain and France. Jack Aubrey, a British naval officer, has many adventures in various ships, mostly those he commands, as well as on land. His friend Stephen is a ship's doctor new when they first meet to the ways of the sea, a scientist devoted to the study of the earth's flora and fauna, and secretly the Admiralty's most valuable intelligence officer in the fight against the terroristic regime of Napoleon Bonaparte.

Maturin's background makes him well suited to intelligence work and allows for quite unconventional cover arrangements. With a Catalan mother and an Irish father in the service of Spain, Maturin grew up speaking Catalan, Spanish, Irish, English, and French, and he intimately knows the topography of Spain and Ireland. A doctor by profession and a natural scientist by vocation, Maturin is well respected—and indeed publishes—in both fields, a situation that provides him with excellent cover for travel to exotic places and for establishing and maintaining contacts worldwide. His appointment as a naval surgeon gives him an official dimension that enhances his nonofficial cover and allows him plausible access to the Admiralty, where he deals with the chief of intelligence.

Temperamentally Maturin is a romantic, fiercely opposing the threat of tyrannical rule posed by France under Napoleon, yet he keeps a coolness and iron discipline that allows him to prevail in dangerous circumstances; it helps that he is a proficient swordsman and an excellent shot. When Maturin needs to, he can kill an enemy, or lie to a friend, but both sicken him. He is curious, always looking to enlarge his knowledge, and he is compassionate about his fellow human beings (as long as they do not serve Napoleon!). His analytic mind serves him well in intelligence work—he has a strong counterintelligence sense about him—and he is happiest when writing "a clear statement of a complex situation." Maturin constantly and imaginatively takes the initiative in collecting and producing intelligence since he has few specific taskings from the Admiralty, which allows him a free hand to act.

Technically proficient, Maturin is expert at codes and secret writing, at breaking and entering, and at "flaps and seals" work for opening envelopes and dispatches. He conducts effective surveillance detection routines. A practitioner of deception—what today we call covert influence operations—he tells Aubrey, "My broadsheets are as effective as your round shot." Being well versed in interrogation techniques allows him to avoid manipulation when it is his turn to be interrogated.

Maturin works for the principle of fighting tyranny rather than for money. He is, in the society of the country he serves and aims to protect, an outsider as a foreigner and a Roman Catholic. All this bolsters his cover; indeed, his own best friend Aubrey is unaware for a long time about Maturin's intelligence work and is shocked when he learns of it. "Now listen, Jack, will you? I am somewhat given to lying: my occasions require it from time to time."

By his looks Maturin is a slight, odd fellow, socially and physically awkward, perfectly at home with the appearance of an eccentric man of science, for that he really is. In reality he is a man of utter discretion and integrity, of keen observation and memory, of immense courage and reliability, with an intense attention to detail and also a calculated risk-taker, a patriot well aware of the shortcomings of the country he serves but totally devoted to the destruction of the evil posed by his country's enemy.

Patrick O'Brian was rightly praised for creating a wonderful world of adventure and human relationships with his Aubrey-Maturin saga, but intelligence officers owe him a special debt for conjuring up in Doctor Stephen Maturin an unlikely but apt literary exemplar for our profession.

Master and Commander and 20 other titles published from 1970 to 2004. Patrick O'Brian. Harper Collins, UK.

The reviewer is a CIA historian, former square rigger sailor, and member of the *Studies in Intelligence Editorial Board. He is the author of The Literary Spy: The Ultimate Source for Quotations on Espionage and Intelligence.*

Stephen Maturin and Intelligence in Patrick O'Brian's Novels

Like all good intelligence officers, Maturin views reports with some skepticism.

Stephen took his short statement out of his pocket and said, "An armament is fitting out in Ferrol, the ships of the San Ildefonso treaty: here is a list of the vessels. Those marked with a cross are ready for sea."... He passed the sheet.

"Perfectly, perfectly," said Sir Joseph, looking at it greedily—he loved a tabulated list, numbers, factual intelligence, rather than the usual vague impressions and hearsay. "Perfect. This corresponds very closely to what we have from Admiral Cochrane."

"Yes," said Stephen. "A little too perfect, maybe."

—*Post Captain* (1972)

He knows that the most committed agent is ideological, not mercenary.

"God help us," cried Jack, gazing at the mass of gold coins lying in a deep curve along the leeward side of the cabin. "What's this?" [Stephen answered,] *"It is technically known as money.... And, I may tell you, every louis, every napoleon, every ducat or doubloon is sound: the French sometimes buy services or intelligence with false coin or paper. That is the kind of thing that gives espionage a bad name."*

"If we pay real money, it is to be presumed we get better intelligence?" said Jack.

"Why, truly, it is much of a muchness: your paid agent and his information are rarely of much consequence. The real jewel, unpurchasable, beyond all price, is the man who hates tyranny as bitterly as I do."

—*The Mauritius Command* (1977)

He convinces others of the need to violate certain principles in fighting the enemy.

Captain Aubrey would do his utmost to deceive an enemy by the use of false colours and false signals, by making him believe that the ship was a harmless merchantman, a neutral, or a compatriot, and by any other ruse that might occur to his fertile mind. All was fair in war: all, except for opening letters, and listening behind doors. If Stephen, on the other hand, could bring Buonaparte one inch nearer to the brink of Hell by opening letters, he would happily violate a whole mail-coach full. "You will read captured des-patches with open glee and exultation," he said, "for you concede that they are public papers. If you value candour, you must therefore admit that any document bearing on the war is also a public paper: you are to rid your mind of these weak prejudices." In his heart Jack remained unconvinced; but Stephen received the letter.

—*Desolation Island* (1978)

Security and discretion is a way of life to him.

"You are a close one, Stephen." "Yes, I suppose I am. I have to be, you know. That is why I am alive."

—HMS Surprise (1973)

Stephen had long practised medicine, a calling in which discretion is often of great importance; but for an even longer time, if time is to be measured by stress, he had been an intelligence-agent, and here discretion was of the very essence, since an unguarded word or step might lead to an agent's death and to the death, the often hideous death, of his friends and the destruction of their cause.

—21: The Final Unfinished Voyage of Jack Aubrey (2004)

He accepts the need for "bulkheads" or compartmentation.

"How is Sir Joseph?" [Maturin] asked [the Admiralty official] when they were alone, referring to his close friend and hierarchical superior the head of Naval Intelligence. "He is physically well," said Philips, "and perhaps a little stouter than when you last saw him: but he is worried. I shall not venture to say what about: you know how cloisonné these matters are with us, if I may use the expression."

"We say bulkheaded in the Navy," observed Stephen.

"Bulkheaded? Thank you, sir, thank you: a far better term."

—The Commodore (1994)

Like every intelligence officer, at his core a collector.

With your permission I will go ashore and gather what information I can: at present I swim in a sea of unknowing.

—21: The Final Unfinished Voyage of Jack Aubrey (2004)

The Spy Who Came in from the Cold

Barry Royden

I first read this book shortly after it was published in the mid 1960s, when I was a young and aspiring CIA case officer. I enjoyed it greatly at the time, which led me to read many of John le Carré's subsequent books through the years. I always found them to be good reading as well as reasonably accurate renditions of the intelligence operational world.

I reread the book recently to remind myself of the story and to see whether I continued to find it a plausible description of the intelligence business. And, with the acceptance of a fair amount of literary license, to which any author is entitled, I concluded that indeed le Carré still presents our world more accurately than most.

The plot is reasonable on many counts. It would be rare, but not impossible, that "Control" (in CIA's case Headquarters) could be running an extremely sensitive and valuable case outside of the normal bureaucracy—i.e., that Leamas as SIS Head of Station/Germany would not be witting of the fact that Mundt, a senior East German Intelligence Official, was an SIS agent. The plan to have Leamas feign disgust with his treatment by the SIS and fall apart to the point where he agrees to sell out to the enemy is also believable (In 1984 former CIA case officer Edward Lee Howard volunteered to the KGB after he had been fired by CIA for cause.)—believable at least to the East Germans. What is not believable is that any intelligence service would run a double agent operation in which a veteran officer is put under the control of the enemy for an open-ended period of time. Such an officer would know too much to allow this to be done. (Example: Vitaliy Yurchenko, the senior KGB officer who defected to CIA in 1985 and then redefected some three months later. It is simply not logical that the KGB would have willingly allowed Yurchenko to be brought to the United States for extensive debriefing under our total control—there is simply too much that he knows after a long career in the KGB that would be of great value to CIA. Le Carré attempts to get around this problem by stipulating that all of Leamas's agents had been eliminated by the East Germans, so he had nothing left of real value to give up, but this is really too much of a stretch to be truly credible.) It is also a stretch that Control would mislead Leamas about who their agent was and then destroy Leamas's cover story by paying his bills in a very easily detectable way, but it makes a great story, which is the whole idea.

Le Carré's description of intelligence operational life as seen through the eyes of a case officer is well grounded. This includes such things as the disgust of the field operative for the "pencil pushers" back in Headquarters (while of course misguided!), and that the agent Karl Riemeck told his girlfriend about his agent role—and Leamas's comment that "all agents lie about something" (very true in my experience). I also think he correctly captures the ethical dilemma of the espionage business—that sometimes you have to deal with/use individuals for whom you have little or no respect (Mundt) but who, because of their access, can provide important service to your government.

I found particularly insightful Leamas's thoughts about "living" the role of the defector—about the need to totally immerse himself in his every waking moment with the idea that he had decided to sell out his service, so that he didn't inadvertently say or do something inconsistent with his cover story. This equates very closely with the kind of training that we have seen as necessary for agents who will be directed to try to penetrate terrorist networks by feigning dedication to their beliefs.

A few inconsistencies are worth pointing out. Riemeck was reportedly directed to recruit subsources, which led to his undoing—no intelligence service worth its salt would risk losing a valuable, well-placed penetration agent by having him aggressively recruit subsources. Also, even the East Germans didn't kill off Western agents on the street when they found them out. At worst they, as the Soviets did, would bring them to trial of some sort.

John le Carré, *The Spy Who Came in from the Cold* (New York: Coward-McCann, 1964), 256 pp.

Barry Royden is a former operations officer and a current member of the *Studies in Intelligence* Editorial Board.

CRESCENT MOON RISING

Noah Rozman

Kerry Collison's latest fictional spinoff of Indonesian political history purports to be "partly a work of fiction, influenced by indisputable historical fact, tempered with not-so-imaginary characters." The book, however, comes across as a lackluster attempt to link a wide range of societal conflicts and acts of terror in Indonesia's recent past to support the questionable assertion that a *dalang*, or puppet master, is secretly controlling events behind the scenes.

Much like Collison's previous books—he has written at least five—which explored similar themes while "explaining events" surrounding Indonesia's invasion of East Timor and the fall of President Soeharto, *Crescent Moon Rising* illustrates the author's depth of experience with intelligence issues and knowledge of Indonesian culture, society and politics—his profile on the books claims he served in Australian military and government intelligence and business. However, Collison's fixation on conspiracy theories involving a confusing nexus of Indonesia's Army Special Forces (KOPASSUS), intelligence agencies, and, in this case, international and domestic terrorists threatens to overwhelm otherwise interesting glimpses into the intrigue and cultural norms that have historically influenced Indonesian policymaking.

Starting with his premise that KOPASSUS officers sponsored the hijacking of Garuda Flight 206 in 1981 to provide an excuse to crack down on outspoken radical clerics at home, Collison portrays Indonesia's military as behind nearly every event, to include the 2002 Bali bombings. Civilian leaders play an astonishingly minor role in this Indonesia—Presidents Wahid, Megawati, and Yudhoyono make what amount to cameo performances, though the publisher's Web site tries to explain away these glaring gaps by referencing Indonesia's "notoriously tough" libel laws, a restriction that does not seem to apply to negative portrayals of Indonesia's Armed Forces. When the military is not undermining harmonious life in Indonesia's outer islands, Collison focuses on outside actors, primarily Dutch, US, and Australian intelligence agencies, but also al-Qaeda. All of these are generally depicted as self-interested and bumbling, although the Dutch and Sin-gaporeans, who are only mentioned twice, come off as slightly more effective than the others.

As a result of this outdated vision of Indonesian politics, and specifically the military's role, which has changed dramatically in the last decade,

Crescent Moon Rising plays directly into long-held Indonesian suspicion that it is only a matter of time before some combination of Western powers again tries to colonize or otherwise dictate to Jakarta. The book also feeds Islamist conspiracy theorists who argue that the Bali bombers—despite their public confessions—could not have carried out the attacks on their own and therefore must have had support from foreign intelligence agencies such as the CIA. In that regard, one wonders if the book was primarily aimed at an Indonesian domestic audience, or whether Collison has bought into a xenophobic view of the outside world that some Indonesians continue to hold.

Kerry Collison, *Crescent Moon Rising—The Bali Bombings* (Sid Harta Publishers, 2008), 440 pp.

Noah Rozman is a member of the Editorial Board of *Studies in Intelligence.*

STORMBREAKER: JAMES BOND FOR A NEW GENERATION

Valerie P.

Some of the best fiction published today is written for younger audiences. But why devote space in an intelligence publication to teen fiction, even if it is about spies? First, what we read when we are younger impresses us deeply and lays the grounding for our developing world-views. Children's fiction reflects the values of society and, in contrast to some literature for adults, still preserves elements of hope for the future. Through books children learn how to live vicariously, how to dare without themselves necessarily jumping off the cliff. Finally, the best teen fiction contains all the elements of good fiction for adults, including compelling plots, fully developed characters, and meaningful themes.

Within genre fiction, spy fiction is particularly ready-made for younger audiences. After all, children easily relate to the world of secrets and special knowledge, apprenticeship, taking on the identity of others (cover), the struggle of the powerless, and overcoming a stronger enemy with wits and guile. In the classic novel for children, *Harriet the Spy*, the title character thrives on uncovering the secret lives of her neighbors and friends, and one need look no further for a current example than the success of Harry Potter, whose arguably most fascinating character is Severus Snape, the classic double agent.

Esteemed British screenwriter Anthony Horowitz taps into the success of the quintessential espionage genre with his teen spy thriller *Storm-breaker*. Like any good thriller, the book begins with a hook: "When the doorbell rings at three in the morning, it's never good news." (Indeed the news is bad: 14-year-old Alex Rider's uncle, who is also his guardian, has been killed in an apparent automobile accident.) From that point on Alex is drawn into the world behind his uncle's "banking" cover and into the business of espionage. Questions surrounding his death propel Alex to investigate what really happened. His investigations lead him into danger when he finds the car peppered with bullet holes and narrowly escapes a grisly death. When he responds to a summons to the "bank" the next day, he's drugged and awakes to find himself in an Elizabethan-style house. There he meets the "bank manager" (in truth the chief executive of Special Operations Division, MI6, Alan Blunt) and Mrs. Jones, head of Special Operations.

Alex learns his uncle was a highly trained field agent for MI6. Alex also learns that his uncle's last—unfinished—operation involved Herod Sayle, a rags-to-riches corporate magnate who has developed a revolutionary computer called Stormbreaker, which uses a new technology. Planning to give away tens of thousands of the advanced computers to British students, Sayle has invited the prime minister himself—a former classmate of Sayle's— in just a few weeks to press the button that will bring all the computers online simultaneously. For his generosity, Sayle is to be granted British citizenship. Convinced that Sayle is too good to be true, MI6 sent Alex's uncle to investigate. He raised a warning, saying the computers mustn't leave the plant in Cornwall, much less be installed in schools around England, but he was killed before he could provide details.

MI6 needs to infiltrate Sayles's secret computer facility at Port Tallon with someone who won't be suspected. Alan Blunt, now that MI6 is Alex's legal guardian, coerces Alex into cooperating. The plan? Insert Alex into Port Tallon in place of the boy who recently won a computer magazine competition to be the first child to use the Stormbreaker. Following a short operations course and armed with spy gadgets but no gun ("we can't give a teenager a gun"), Alex deploys. Inside Port Tallon Alex follows in the steps of his uncle, discovering clues and encountering the requisite car chases, plane crashes, and other near misses that keep readers turning pages. He discovers the deadly secret and finds a way to prevent the release of a computer virus unlike anything previously concocted.

In creating the Alex Rider series with *Stormbreaker*, the first of seven books to date, Horowitz has given more than a few nods to James Bond. Like Bond, Alex is an orphan, an often used situation in children's fiction because it gives the child the least adult supervision and allows the character—freed from parental bonds—to craft his or her own identity. When the MI6 CEO tells Alex his uncle was a spy, Alex even asks, "You mean…like James Bond?" Alex earns the nickname "Double O Nothing" during his foray into operations training, where he learns map reading, radio communication, first aid, and basic survival. Also like James Bond, he receives spy tools from a shuffling, yet Q-like character, Mr. Smithers. Alex's spy kit includes a motorized yo-yo strung with nylon strong enough to hoist 200 pounds; zit cream that, when applied to metal, dissolves it; and a hand-held gaming device installed with four "games" that can also be programmed to communicate covertly with MI6, take x-rays, detect listening devices, and set off smoke bombs for diversionary tactics.

Alex's personality also suggests the 007 influence. Alex has a habit of speaking his mind in a clever way that puts him into hot water. He's familiar with understatement and irony. He finds killing distasteful even when he kills out of necessity. The deaths of the villains often occur indirectly—for example, in a plane crash caused by Alex's use of the smoke bombs. Espionage isn't necessarily about war and killing, but sport and gamesmanship—and winning.

Horowitz's non-Anglo-Saxon villains closely follow British spy genre expectations. Key nemesis Sayle (remember his first name? Herod) is Egyptian born and, as is revealed, suffered ridicule and physical abuse from other British school boys, setting up deep-seated psychological issues that help explain his mad scientist revenge scheme. Like other villains with dangerous pets, Sayle keeps a giant Portuguese man-of-war in a huge aquarium in his office because it reminds him of himself. "It's an outsider," he tells Alex, which "drifts on its own, ignored by the other fish. It is silent and yet it demands respect." (Any spy fan instinctively knows the sea creature will play a grisly role later in the plot.) The villain's henchman, Mr. Grin, reminds readers of Jaws from the Bond films *The Spy Who Loved Me and Moonraker*. The German-accented female accomplice Fraulein Nadia Vole (Vole being another perfect villain name) fits the genre stereotype of female villains who are not just evil but untrustworthy. (On the flip side, Horowitz's female protagonist Mrs. Jones, head of special operations, follows well-trodden expectations for the female operative at the Headquarters level—no first name, somewhat unfeminine and cold, and definitely devoid of a sexual nature, though Alex briefly wonders if she is a mother with a son about his age.)

In another nod toward the genre, Horowitz's Russian contract assassin Yassen Greorovich comes with a code of ethics of sorts, an honor code among those in the game. Though he killed Alex's uncle, Yassen, in a twist at the end actually saves Alex from death at Sayle's hand. When Alex asks the Russian if he is going to kill him, Yassen responds that he "has no instructions" regarding Alex. He refuses to kill Alex even when Alex says he "one day" will kill him. Horowitz predictably but compellingly sets up another confrontation for a future installment—and he correctly shades the sometimes ambivalent moral dilemmas inherent in the best spy fiction. At the end Yassen raises his hand. "A gesture of friendship? A salute? Alex raised his hand." All that's missing in this Bond-like, fast-paced spy thriller is a teen love interest for Alex. Perhaps that is yet to come too.

Stormbreaker is a good, fun read not just for teens but for any aficionado of genre spy fiction who yearns for the (very) familiar plot line but still wants spell-binding entertainment without having to think too hard. Not only does it contain the best elements of the genre, at heart the novel also incorporates the essential elements of the hero's journey—the call to adventure, the helper with his magic tools (Mr. Smithers' spy kit), crossing the threshold, the trials, the woman as temptress, and the often treacherous return journey. By following in the steps of the hero's journey, Horowitz taps into a fundamental yearning in all of us—adults and children—for story and meaning.

Stormbreaker, An Alex Rider Adventure
(Anthony Horowitz. Scholastic Books, 2006)

Valerie P. is a senior attorney in the CIA's Office of General Counsel and a member of the *Studies in Intelligence* Editorial Board. She is the author of a forthcoming novel for teenagers.

ROGUE'S MARCH

James M. Burridge

Samuel Hamrick, writing as W. T. Tyler, wrote a small but distinguished body of fiction after his retirement from the US Foreign Service in 1980 until his death in 2008. This novel is set in Zaire in 1970 and tells the story of a violent military coup. The main character is Andy Reddish, the CIA deputy station chief (D/COS), who is nearing the end of a four-year tour in Kinshasa. During the coup and its immediate aftermath the COS is absent, taking time to attend a conference of Africa chiefs of station at Headquarters and to take personal leave.

Reddish is very close to the old and corrupt president and is widely known in Kinshasa as a CIA man who gets things done—such as providing M-16s unavailable through official US government channels. Although it isn't explained why the previous US ambassador gave Reddish free rein, the current ambassador was assured by the director of central intelligence (DCI) himself that "the old ways of dealing with the president were over, that there would be no more midnight visits to the palace." Bondurant, the ambassador, doesn't trust Reddish and accuses him of continuing to use his assets to buy influence rather than intelligence.

The ostensible reason for the coup is an armed uprising by a minor Marxist party, but most of the novel concerns Reddish's quest to determine the real relationship between the two events and the source of the rebels' arms. Reddish learns that the weapons came from the army, which intended to provoke the rebellion. He also discovers that the US defense attaché inadvertently facilitated the provocation by sharing with the army reporting on an arms shipment transiting Zaire en route to the Popular Movement for the Liberation of Angola. Reddish reports all this to Headquarters.

Although Hamrick has been compared to Graham Greene and John le Carré, Reddish—unlike Greene's protagonists and George Smiley—doesn't agonize over moral judgments. Although he has more empathy for and knowledge of the Congolese people than any of his embassy or station colleagues, he sees the governance situation as hopeless and doesn't view the new government as any less legitimate than the deposed one. Bondurant tells him, "A few hundred bewildered, untrained Africans are given guns and then shot down to justify someone's obscene grab for power, and you tell me the idea wasn't at fault, just the execution. You amaze me, Reddish—absolutely bewilder and amaze me." Reddish's is a cold quest for abstract truth.

The ending is decidedly anticlimactic. The COS and Bondurant are unhappy with Reddish's reporting because they (and Washington, preoccupied with Vietnam) have come to believe they can live with the military regime, and the reports might muddy the waters. But the reporting has no impact. His replacement arrives, and Reddish leaves Zaire with no follow-on assignment.

Hamrick resisted the label "spy novelist," and *Rogue's March* lacks many of the elements associated with contemporary espionage novels—elaborate descriptions of tradecraft, double-agent betrayals, and introspection and angst about the morality of the whole enterprise. He creates a compelling picture of the political and social environment in which Reddish conducts his work—his necessary if resented relationships with the ambassador, the deputy chief of mission, the political counselor, and the defense attache. During the crisis he interacts with them more than he does with the case officers whom he directs, in part because most of the station's assets are killed or jailed by the army. Far from the singleton lone-wolf operator of so many espionage novels, Hamrick's protagonist functions in a bureaucratic context that will resonate with anyone who has worked in the real world of American intelligence.

W. T. Tyler, *Rogue's March* (New York: Harper and Row, 1982), 345 pp.

James Burridge managed the Intelligence Community's support to multinational organizations from 1998 until his retirement in 2007. He has also served in the Bureau of Intelligence and Research in the State Department. He now works as an independent contractor at CIA. He is author of three articles for *Studies.*

In: Intelligence in Contemporary Media
Editors: Frances M. Walsch
ISBN: 978-1-61122-651-5

Chapter 3

INTELLIGENCE IN BOOK AND FILM

THE HUNT FOR RED OCTOBER: THE TECHNO-ESPIONAGE PROTOTYPE?

Bill Hadley

As a sailor and historian, I have always been fascinated by Tom Clancy's book *The Hunt for Red October* because its detailed technical jargon roughly parallels advances in sailing that I have experienced. The ability of the Los Angeles Class attack submarine, the USS *Dallas,* to chase the *Red October* through dangerous submarine canyons in their odyssey across the Atlantic by use of gravity gradiometry (more on this later) is one example. It reminds me of the development of electronic chart plotting that allows sailors to see where they are regardless of fog or other adverse weather. Such a system has saved me many times from disaster, including a midnight entrance several years ago in heavy fog through a rockbound jetty to Great Salt Pond at Block Island in Rhode Island Sound.

- **Movie Release Date:** March 1990
- **Director:** John McTiernan *(Predator, Die Hard)*
- **Film Company:** Mace Neufeld Productions; Nina Saxon Film Design; Paramount Pictures
- **Screenwriters:** Larry Fergu-son and Donald Stewart
- **Domestic Gross:** $120.7 million
- **Foreign Gross:** $78.5 million

Most readers will know the *Red October* story. The Lithuanian-born Marko Ramius (Sean Connery) is captain of the Soviets' newest Typhoon-class ballistic missile submarine, *Red October*, equipped with a revolutionary silent propulsion system, nicknamed "Caterpillar Drive." The system is supposed to make sonar detection next to impossible. Ramius, disillusioned with the Soviet system and convinced that Moscow aims to use the new submarine as a first-strike weapon against the United States, concocts a plot to defect and hand over the submarine to US authorities. The Soviet North Fleet gives chase to stop him. The US Navy at first believes he is a renegade bent on destruction. But the investigations of CIA analyst Jack Ryan (Alec Baldwin) lead US National Security Advisor Dr. Jeffrey Pelt (Richard Jordan) to recognize Ramius's intended defection, and a plan is developed to help him to do so, all the while fooling the Soviets into believing that the submarine sank and was not recovered.

The novel reportedly was inspired by two real-life Soviet naval defections. The first was the successful 1961 defection of Lithuanian-born Captain Jonas Pleskys, who sailed his Soviet submarine tender to Sweden. The second was the ill-fated 1975 defection attempt by Political Officer Valery Sablin of the frigate *Storozhevoy (Vigilant)*. Soviet aircraft disabled his ship and Soviet marine commandos boarded and recovered it. Sablin was executed by firing squad.

The Hunt for Red October was the first work of fiction ever published by the Naval Institute Press, in 1984. Moviemakers avoided it for a long time because they regarded the story as too complicated. Ultimately producer Mace Neufeld took on the project. Several senior US naval officers—convinced the movie would do for submariners what *Top Gun* had done to boost the image of US Navy jet fighter pilots—provided unprecedented access to their submarines and training in submarine steering for Con-nery, Baldwin, and Scott Glenn (commander of the *Dallas*). The navy even allowed the use of its subs in the film. The USS *Houston* (in the role of the *Dallas)* reportedly made more than 40 emergency surfacing "blows" for rehearsal and for the cameras. Eventually, the Navy involved the aircraft carrier USS *Enterprise,* helicopters, two frigates, and a dry-dock crew. Most reviewers panned the film, but it grossed $17 million during its first weekend in March 1990 (more than half its budget) and eventually earned about $200 million.

Clancy explained the fate of the *Red October* in his next novel, *The Cardinal of the Kremlin,* which revealed that the United States had reverse-engineered its propulsion system, stripped it of its technology, and sunk it in a deep trench off Puerto Rico. The film became controversial in some official

US circles as rumors spread that the movie script revealed a billion-dollar black project undertaken for the Navy by Bell Aerospace, which involved the development of a so-called full-tensor gravity gradiometer. This technology—widely used today in oil, mining, and other industries—facilitated production of excellent images of the seabed and marine mountains for use by the Navy's Trident missile submarines. It is alleged to have been deployed on only a few Ohio-class Trident submarines after first being developed in the early 1970s but, tellingly, was declassified a few months after *Red October* was released.

The Hunt for Red October is a major link in a long chain of spy fiction that arose before World War I in the aftermath of the infamous Dreyfus affair in France. Many of the pre-World War I spy novels featured themes on the competition for empire between Britain and Germany. A good early example was Erskine Childers's famous novel, *The Riddle of the Sands* (1903), which focused on a German plot to stage a naval invasion of Britain from the Frisian Islands (later made into a minor movie classic featuring Michael York). Other examples include the novels of William le Queux, such as *The Great War in England in 1897, Spies of the Kaiser,* and *The Invasion of 1910,* the latter a major bestseller. Another widely read spy fiction writer before the Great War was Edward Phillips Oppen-heim, the author of more than 150 books, most of them spy thrillers, beginning with *Mysterious Mr. Sabin* in 1898.

Many others followed between World Wars I and II, including for the first time novels by former intelligence officers such as W. Somerset Maugham (*Ashenden*, the story of a sophisticated and aloof spy in which Maugham drew on his World War I experiences as a spy), Sir Edward Compton Mackenzie, who wrote four books drawing on his work for British intelligence in the eastern Mediterranean during WW I. World War II prompted even more, including Helen MacInnes's first espionage novel, *Above Suspicion,* which was followed by *Assignment in Brittany*. Many books of this period were made into films.

The Cold War led to a huge expansion in spy literature. Graham Greene drew on his personal experiences with British intelligence to create a number of leftwing, anti-imperialist spy novels, including *The Quiet American* (1955) set in southeast Asia, *A Burnt-out Case* (1961) about the Belgian Congo, *Our Man in Havana* (1958), and *The Honorary Consul* (1973) set in Argentina. Of course, in this period Ian Fleming exploded onto the scene with his James Bond novels—and over 20 movie spinoffs. Others appeared as well, including John le Carré and Len Deighton. American novelists also began to establish themselves, including Edward Aarons with his *Assignment* series of 42 novels about CIA agent Sam Durell in the mid-1970s; Donald Hamilton, who created the undercover agent Matt Helm; and Robert Ludlum, the author of 25 novels that sold nearly 300 million copies including the now famous Bourne series.

Given this rich heritage of espionage fiction, we cannot say accurately that *Red October* initiated the era of the techno-espionage thriller, however, Clancy's work must be given credit for popularizing the genre. As one reviewer of *Red October* noted—in comments that could be applied to many of the techno/espionage genre—"The plot involves us in the clever and deceptive game being played, made even more fascinating because of the high technology involved." This characteristic clearly separates more traditional spy novels from their thrilling, high-tech competitors of the present.

Red October added a high gloss of popularity to this genre, but it was not the first successful such creature. The many James Bond novels and movies hint at this genre with "Q's" many ingenious briefcase devices and hidden weapons and other high-tech features embedded in Bond's various cars. Some of the James Bond movies go even further into sci-fi technology.

A case for a frontrunner can be made for Craig Thomas's *Firefox* about a fictional Soviet MiG aircraft with advanced capabilities, including a form of stealth technology, a design allowing it to fly at Mach 5, and weapons controlled by the pilot's thought impulses. The novel was published in 1977, fully seven years before *Red October*. Warner Brothers made it into a movie released in 1982 (nearly a decade before the film version of *Red October*), with Clint Eastwood as the director, producer, and pilot engaged to steal it.

We have been lucky to have such a galaxy of talented writers and moviemakers. We are certain to be reading many more books in this grand tradition and enjoying their movie versions in the years to come as the ever-expanding world of high technology continues to provide limitless material.

Bill Hadley is a retired CIA senior officer who specialized in the Middle East and Africa. He is currently a Senior Technical Director for Global Analysis, BAE Systems. He is a lifelong, avid sailor who owns a 36-foot ketch named *Amazing Grace*.

THE KITE RUNNER

Elizabeth Darcy

I loved the 2003 international bestseller *The Kite Runner* by Khaled Hosseini. The novel brought to life tumultuous periods in Afghanistan beginning in the 1970s through to the reign of the Taliban in a compelling, riveting way as it followed the lives of two young Afghan boys. I have to admit that I wanted to like the movie far more than I actually did. The movie, while a good representation of the novel, doesn't capture the rich narrative, the drama, or the emotional upheaval of the main characters in the story as they live through and survive (most of them anyway) Afghanistan's modern history. Still, while I preferred the novel, I would recommend the movie to anyone interested in Afghanistan, its people, or merely in a good piece of fiction. If you don't have a day to devote to the novel—which you most likely won't be able to put down (or want to stop listening to in its audio version)—then I highly recommend the movie as the next best thing.

The Movie

- **Release Date:**December 2007
- **Director:**Marc Forster (Monster's Ball, Quantum of Solace)
- **Film Company:**Dreamworks SKG; MacDonald/Parkes Productions; Neal Street Productions; Participant Productions; Sidney Kimmel Entertainment; Wonderland Films
- **Screenwriters:** David Benioff and Khaled Hosseini (novel)
- **Domestic Gross:** $15.8 million
- **Foreign Gross:** $57.5 million

Ever since living and working in Pakistan in the late 1990s, I have been fascinated—and horrified—by the history of Afghanistan and its people. I am sure anyone with experience in the region is well aware of the modern history of the Afghan people. It is a complex, tragic story that involves the clash of civilizations, the "Great Game" (traditionally referring to the secret war between Tsarist Russia and Victorian Britain for control over central Asia), and multiple invasions, in short, a story of survival, an amazing, enduring characteristic of the Afghan people.

The Kite Runner is emotional tale of two Afghan boys and their families, starting in the relatively chaos-free years before the Soviet invasion and concluding in the middle of the Taliban's theocratic lockdown. The story begins in 1978 with Amir and Hassan playing in the streets of Kabul. Hassan and his father, Ali, work as servants for Amir and his father, Baba. Amir and Hassan are best friends and team up to win Kabul's annual kite flying competition. The kite flying competition is a major cultural event, but in this story it is capped by a savage act and Amir's cowardice in confronting that act. The event immediately changes the way the boys see each other and their world. Soon thereafter, the Russians invade Afghanistan, and Amir and his father escape to America where they begin to make a new life for themselves. Although he adjusts to life in America, Amir is haunted by his cowardice on that day of the kite flying competition, and it is clear he is consumed by regret. Later, given an opportunity to try to make things right—even if it means risking his life—Amir returns to Afghanistan to rescue Hassan's recently orphaned son, who we later discover is also Amir's nephew. He chooses to do so, and this is where the story gets really interesting.

Where the book leaves you to imagine Afghanistan's setting, the scenery in the movie is breathtaking and surreal. Most of the movie was filmed in Western China, but portions were also filmed in Kabul. The Tal-iban protested the filming in Kabul, given their religious prohibitions against movies in general, as well as their objections to the content and to their negative portrayal. The Taliban also threatened the young actors—all of whom were Afghans—and they and their families eventually had to be settled outside of Afghanistan by the movie production company. Although the book apparently has yet to be translated into Pashtu, the Taliban's hostility to it goes to show that the pen can indeed be mightier than the sword.

Elizabeth Darcy is the pen name for a CIA case officer with over 13 years experience in the National Clandestine Service. She has served in several locations in South Asia. She currently works in the Intelligence Programs Directorate in the Executive Office of the President at the National Security Council.

The constant sense of loss for the innocence of Amir's childhood mirrors the sense of loss and suffering in Afghanistan, first to the Soviets and then to the Taliban. While the movie's ending is somewhat predictable, it remains poignant against the backdrop of the horrors being imposed on the Afghan people by the Taliban. Amir is not necessarily meant to stand in as the hero of or for the Afghan people; no doubt Hosseini chose an ordinary character around whom to build his story to represent the will to survive of the ordinary Afghan people.

Can *The Kite Runner* help prepare intelligence officers for the uniqueness of their work on the subcontinent? I get that question a lot, and I think it can help, at least partially. While neither the novel nor the movie captures the essence of intelligence work in the region (I would recommend Peter Hopkirk's classic nonfiction study, *The Great Game: On Secret Service in High Asia,* for that purpose), Hosseini's book is a solid primer on the people and modern history of Afghanistan. As any intelligence officer will tell you, to successfully work against a particular target, you first must understand the target. By reading the novel, and to a lesser extent by watching the movie, an intelligence officer new to the region will be effectively and memorably introduced to both the people (and tribes) of Afghanistan as well as modern Afghan history.

In: Intelligence in Contemporary Media
Editors: Frances M. Walsch
ISBN: 978-1-61122-651-5

Chapter 4

INTELLIGENCE IN FILM AND TELEVISION

ONE DAY IN SEPTEMBER AND *MUNICH:* ENDURING QUESTIONS, INDELIBLE IMAGES

Dr. Daniel Tsao, M.D.

My youthful memories of the massacre of 11 Israeli athletes at the 1972 Munich Olympics by members of the Palestinian group Black September were instantly and chillingly revived by two films centered on that event: Kevin MacDonald's 1999 documentary *One Day in September* and Steven Spielberg's *Munich*, released in 2005.[1] Although not without flaws (which were well-publicized, especially for *Munich*), both films compellingly raise questions for those in the field of antiterror-ism and intelligence.

One Day, which received an Oscar for best documentary for 1999, painstakingly details the events of that tragic day, from the terrorists climbing the walls of the Olympic compound to graphic images of the athletes' bodies following the botched rescue attempt. The heartbreaking personal stories of the athletes, including interviews with surviving family members, leave no doubt—especially to those of us directly involved in fighting terrorism—about the stakes involved but also raise inevitable questions: Could it have been prevented? Could it have ended differently? It is in this realm that the film's

[1] Nine members of Black September, a PLO entity, took part in the hostage taking. Six died in a failed attempt to save the hostages. The remaining three were held by the Germans but were released in exchange for a hijacked Lufthansa airliner.

subjectivity is clearly apparent in its negative portrayal of the conduct of West German authorities in confronting the crisis. Will the lessons of that experience (the film does briefly mention West Germany's subsequent development of the renowned GSG 9 counterterrorist unit) be lost, despite our more recent traumas, on future generations?

- ***One Day in September*** (documentary)
- **Release Date:** October 1999
- **Director:** Kevin MacDonald *(State of Play, Last King of Scotland)*
- **Film Company:** Passion Pictures; Arthur Cohn Productions; BBC; British Screen Production (collaboration); Dan Valley Film AG; European Co-production Fund (collaboration). Sony Classics (US theatrical distributor)
- **Screenwriters:** None listed
- **Domestic Gross:** $156,818
- **Foreign Gross:** Not available
- ***Munich***
- **Release Date:** January 2006
- **Director:** Steven Spielberg (*Schindler's List, Saving Private Ryan; Minority Report*)
- **Film Company:** Dreamworks SKG; Universal Pictures
- **Screenwriters:** Tony Kushner and Eric Roth
- **Domestic Gross:** $47.4 million
- **Foreign Gross:** $82.9 million

Munich, which was nominated for the Oscar for best picture, clearly falls into the category of Spielberg's films with thought-provoking messages (in contrast to his commercially successful, crowd-pleasing blockbusters). "Inspired by real events," the film focuses on an Israeli assassination squad dispatched to avenge the massacre by killing those who allegedly planned the attack. It ambitiously attempts to answer a question that *One Day* was criticized by some for not attempting to answer, "What motivated the terrorists to act?" In one scene, a Palestinian terrorist, debating (unwittingly) with the head of the Israeli assassination squad, movingly speaks of the need for "home," no matter how distant, whether measured in miles or generations. Spielberg, who vigorously denies charges that the movie equated Israeli actions with those of Palestinian terrorists, acknowledges that he meant to raise the question, "Do efforts to counter terrorism produce unintended consequences?" This inevitably leads to a larger, personalized question, applicable to all of us in the business of intelligence: although our work is surely not humanitarian in nature, can we preserve our sense of humanity?

Among the many powerful images shown in both movies, two stand out in my mind. One, shown in both films, is newsreel footage of the three surviving terrorists giving a press conference after being released (under questionable circumstances) less than two months after the massacre. Their smug and smiling faces starkly remind us that no film can adequately answer the timeless question: What is the nature of evil? The other image, not visually depicted, demonstrates that as imaginative as films are, fact is often truly stranger than fiction. It is that of Mossad chief Zvi Zamir, standing on a German airfield one generation removed from the end of World War II, watching the disastrous rescue attempt unfold and shouting in Arabic to the terrorists to surrender, only to duck under a return of gunfire that was the answer to his plea.

Daniel Tsao, MD, is a senior psychiatrist in CIA Medical and Psychological Assessments.

THE SIEGE

Eric Heller

For most of the American public before September 11th, the world of counterterrorism operations and their implications were not the subject of robust public discourse. Although some people were familiar, from afar, with the Middle East terrorism of the 1970s and 1980s, not until the 1993 attack on the World Trade Center, the 1995 bombing in Oklahoma City, and the 1996 attack on the Khobar Towers did US society more broadly feel the impact of terrorism. The 1998 film *The Siege,* while taking the usual Hollywood liberties, proved prescient in its presentation of the very real issues the United States would face squarely after September 11th.

The Siege tells the story of the US reaction to a wave of terrorist attacks by a shadowy network of Islamic extremists in New York City in retaliation for the US detention of an influential extremist sheikh in the Middle East. As a result of continued attacks, the president declares martial law in New York.

The declaration pits the efforts of a CIA officer and an FBI task force against US military operations in the city to end the siege. In this story line, watchers cannot help but be reminded of the Khobar Towers attack and the detention of the "Blind Sheikh," Omar Abd al-Rahman, who was convicted in the United States for his role in the 1993 World Trade Center bombing and who has often been the subject of release demands from extremist groups.

- **Release Date:** November 1998
- **Director:** Edward Zwick (Glory, The Last Samurai)
- **Film Company:** Bedford Falls Production; 20th Century Fox
- **Screenwriters:** Lawrence Wright (story and screenplay); Edward Zwick (screenplay)
- **Domestic Gross:** $41 million
- **Foreign Gross:** $57 million

More broadly, *The Siege* depicts the debate over the question of whether terrorist acts should be considered acts of war to be handled by the military, or illegal acts to be confronted by law enforcement authorities and subject to methodical, investigative procedures. The army officer eventually put in charge of military operations in New York City, Major General Deveraux, warns senior US policymakers not to use the military in a US city— "the Army is a broad sword, not a scalpel." The movie also includes the Hollywood-required dynamics of uncooperative, stove-piped FBI, CIA, and military people not sharing information with one another—if not deliberately working at odds with one another.

The movie goes beyond that cliché, however, to bring to the viewer the idea that the terrorist threat represents a new paradigm in conflict, that of a nebulous, loosely connected cellular network; that it includes suicidal terrorism; that it involves Islamic extremists; and that it presents a highly complex analytical problem. These issues may be commonly accepted today, but they were new to many Americans in 1998.

Those watching in 2009 will find that the movie foresaw other issues that are widely debated now. Despite Devereaux's warning, the army is called into action and begins mass roundups of Arab males, moving them into a makeshift detention center in a converted sports stadium. *The Siege* plays out the media fervor surrounding the events, shows a local community's paralysis in the face of the attacks, and highlights a spike in hate crimes against Arab communities. One of the defining scenes of the movie, near its climax, depicts an Arab man stripped naked in the middle of a room, interrogated by a woman, who later debates with MG Deveraux the efficacy of various forms of torture to gain information from a subject who might be described as a "ticking time bomb." The scene ends with the sounds of torture and an execution as MG Deveraux—playing lead interrogator, torturer, and executioner—proclaims, "the time has come for one man to suffer to save hundreds of lives." Throughout the scene, key characters debate how terrorists and terrorist threats ought to be addressed and at what point the United States starts to sacrifice what it stands for.

The Siege is by no means without hyperbolic flair: CIA officers entangled in romantic relations with clandestine agents; military officers executing US citizens; supervisory FBI special agents storming classrooms held by terrorists; CIA officers making amends for a failed covert operation by facilitating the movement of terrorists into the United States; and full frontal military assaults on auto body shops in Brooklyn are but the tip of the iceberg. This, of course, says nothing about the liberties taken with US bureaucratic politics, timelines, and resources.

Nevertheless, *The Siege* lays clear the issues that US and international policymakers and intelligence leaders have had to consider. Ahead of its time, the movie confronts in broad terms complex issues—military operations in the US and posse commitatus, interrogations, torture, detention policies, racial profiling, the utility of military versus law enforcement efforts in counterterrorism, in addition to other ethical and legal issues that Americans would later face. Films like *The Siege* deserve our attention, no matter how hyperbolically portrayed. They tell us how some people think about these issues well before they are in vogue.

Eric Heller is a Department of Defense intelligence officer who has specialized in counterterrorism for the past six years. He is a member of the Editorial Board of *Studies in Intelligence* and served multiple combat tours since 9/11.

9/11 Documentary

Dennis C. Wilder

Even with the passage of time, the documentary film made by two French filmmakers of the 9/11 attack on the World Trade Center is extremely hard to watch. My heart still aches to see the pictures of brave public servants of the New York police and fire departments entering the towers immediately after the airplane strikes without any idea that they themselves would soon become victims. At points, such as when the grey ash has fallen covering the cars and streets of New York, it has the look of a Hollywood thriller, and you almost expect to see Bruce Willis or Batman show up to save the day. But no Hollywood superheroes showed up that day in New York City—just run-of-the mill civil servants, who barely had time that morning for their first cup of coffee, before being thrown into the most momentous day of their generation.

- **Release Date:** March 2002
- **Director:** James Hanlon; Rob Klug; Gédéon Naudet; Jules Naudet
- **Film Company:** CBS; Goldfish Pictures; Reveille Production; Silverstar Productions
- **Writers:** Tom Forman; Greg Kandra
- **Domestic Gross:** Had the largest audience ever for a TV documentary-39 million viewers-when broadcast on CBS on 10 March 2002.
- **Foreign Gross:** NA

I arrived at CIA Headquarters that morning just after the first attack on the towers to find employees standing around in shock staring at their televisions sets. Rumors were flying—the State Department and Pentagon had been hit, there were more planes on the way, and our Headquarters was one of the primary targets. We had known for a long time that terrorists had been plotting to attack CIA headquarters. After the second strike, the decision was made to evacuate CIA buildings and to send most of the employees home. Office-level managers, such as myself, gathered together outside the Headquarters building to await instructions from Director Tenet. As we began to gather, I remember thinking as I looked up into the clear blue sky that this was in a small way what it must have been like to be standing in London during World War II waiting for the deadly and unpredictable V-1 and V-2 rocket attacks.

While the heroes of lower Manhattan were captured on film that morning, I remember many other unheralded acts of courage that unfolded that day in front of my eyes at Headquarters. In the Office of Medical Services, senior doctors stayed in their offices. The deputy director of OMS firmly sent most of his subordinate medical staff members, including my wife, out of the buildings, but remained behind to provide leadership to the skeleton medical team that was assisting in the evacuation, providing support to other stay-behind essential personnel, and in the worst possible scenario, who would function as "first responders" to the wounded from inside the blast area should the buildings be hit and the doctors themselves survive.

Steve Kappes, then the senior officer in the Directorate of Operations (now NCS) for Counterintelligence, stood guard like the steely-eyed United States Marine that he is among the crowd of office directors outside of the outlying building on the CIA campus that had become the temporary command center for the DCI and his immediate staff. Steve had a tight look on his face and when I asked him who could have launched such an unbelievable attack, he quietly looked me in the eye and said: "It has to be bin Laden." An order finally came from Tenet for those of us in positions of authority to decide on one employee from each line unit to send back into the building to ensure that we maintained a capability to answer any questions asked by senior policymakers. I remember trying to decide who I would ask to take up this potentially deadly duty. I decided that the only logical choice was to ask a fit, young, unmarried officer to return to his post. When I spoke with Jeffrey Waggett about what was required, I heard fear in his voice but also saw in his eyes profound commitment to duty and steady resolve.

We should all be grateful to the French documentary crew for capturing for history the acts of heroism in New York City that morning that stand so profoundly as a rebuke to those who so callously and with such cruel calculation took the lives of so many innocents. We should be grateful to the French filmmakers for performing their own form of duty by staying on-site and recording the terrible events of that day. They showed their own courage, dedication, and professionalism as filmmakers in their actions. Because of them, future generations will be able to see images from this great struggle with evil—unvarnished and unretouched—just as we have been privileged to see those searing images taken of soldiers struggling to gain a foothold on Omaha Beach on D-Day in 1944.

But for me, the image that I will carry the rest of my life is of that young, scared but steadfast analyst making his lonely way back into the empty halls of CIA Headquarters, with no idea of whether he would see another day.

Dennis C. Wilder, a career analyst and senior manager at CIA, is currently on academic sabbatical at the John L. Thornton China Center at the Brook-ings Institution.

Hamburg Cell

Senior New Zealand Security Intelligence Service Case Officer

The *Hamburg Cell* is a dramatization of the events leading up to the 11 September attacks. The film is directed by British director Antonia Bird and is based on personal interviews, unpublished correspondence, and the official *9/11 Commission Report.* For the most part, the film focuses on the life of one of the hijackers, Ziad Jarrah, although viewers are also given an insight into the life of fellow hijacker Mohamed Atta and, to a lesser extent, recruiter and talent spotter Ramzi bin al-Shibh.

- **Release Date:** January 2005 (Europe)
- **Director:** Antonia Bird
- **Film Company:** Mintorn Television; Inner Circle Pictures
- **Screenwriters:** Ronan Bennett; Alice Perman
- **Domestic Gross:** NA
- **Foreign Gross:** NA
- **Miscellaneous:** DVD released in November 2006, but otherwise it was apparently not commercially shown in the United States.

Rather than focus on the attack itself, the film explores the radicalization of Jarrah in Germany. The film begins with his recruitment into a Muslim prayer group, moves quickly to his exposure to their extremist rhetoric, and culminates in his training in Afghanistan, where he is chosen to participate in the 11 September attacks.

While Jarrah's journey begins with a desire to become a better Muslim, his exposure to the thoughts, beliefs, and opinions of the fundamental Muslims in his prayer group causes him to question his belief system and the way he looks at the world. As the group shuts out those who attempt to speak out against them, their self-reinforcing behavior has a radicalizing effect on Jarrah, and Jarrah finds himself espousing the same rhetoric as his brothers, much to the dislike of his more moderate and progressive girlfriend, Aysel. While Atta is portrayed as the stern hardcore fundamentalist from the beginning, the audience is given a glimpse into Atta's background later in the film and the pressures brought to bear on him by a family of overachievers.

By focusing more on the development of Jarrah (as opposed to the logistical planning and execution of the attacks), viewers are exposed more to the impact of Jarrah's radicalization, particularly on Jarrah's loved ones. While the film does well in illustrating the camaraderie Jarrah has with his fellow prayer group attendees, Jarrah's radicalization happens a little too easily, and he appears to offer little, if any, resistance to the more extreme views of his "brothers." The only other criticism of the film is that the portrayal of Jarrah is a little wooden, with stronger performances by the actors playing Atta and al-Shibh. Notwithstanding, the film is a worthwhile watch.

BAGHDAD ER—THE 86TH COMBAT SUPPORT HOSPITAL IN IRAQ, A DOCUMENTARY

Dr. John Elliott, MD

One day during my time in Baghdad revolved around trying to scrounge medical supplies for a young Iraqi man who had suffered third degree burns over 75 percent of his body when a bomb went off at a shop he was passing by. He would have been unlikely to make it even at a top-tier burn center, but in his current circumstances—just discharged from the combat support hospital (CSH) to get ongoing care locally—his prognosis was even poorer. But the doctors had done the right thing, and in the grim world of high-volume emergency triage, civilian or military, such hard decisions are made dozens of times a day.

- **Release Date:** 2006
- **Director:** Jon Alpert; Matthew O'Niell
- **Film Company:** HBO Documentary
- **Domestic Gross:** NA
- **Foreign Gross:** NA

Later that night, I brought in another Iraqi, one who had been shot through the leg. He was fortunate; the rounds had punched through his thigh and only damaged soft tissue. The ER was quiet when we brought him in. The staff had just finishing cleaning up the last of the gore from a prior case. Then the quiet fell away as a stretcher bearing a young American soldier was pushed through the doors: Sniper victim; gunshot wound to the head; bandage around the left temple soaked through with blood; heart beating; lungs working; neurological function marginal. Medical orders were shouted. Nurses and doctors were working on the patient as the gurney rolled in. The medic who had retrieved the soldier from the battlefield briefed the CSH staff. The cacophony of a busy ER rang around the tiled walls and the work surged on.

In many respects, the experience at the CSH is the same as at any major urban ER. There are gunshots and car wrecks and death and dying just like in the civilian world. In this the film does a good job of capturing the real bloody mess that ER work often can be. The film also provides a good feel for the pace of work in pre-2007 Baghdad at the height of the carnage. It offers good general perspectives of medical staffs—the gallows humor, the anger at the enemy for causing whatever injury was next to be treated, the appreciation of down time.

The film has a primary failing, although it is probably inescapable. Medical films (documentaries or not) invariably use editing and background music to create a sense of drama that is just not there on the ground. The men and women working the ER approach the work without the shows of emotions, hand-wringing, and existential angst that is typical in contemporary medical dramas. Training and experience have engendered in medical professionals a capacity for clinical detachment that is required to effectively intervene in harrowing situations and to save lives.

To be fair, people watch films (even documentaries) to be entertained and adding to the entertainment value with compelling music and dramatic editing is just what filmmakers do…at least if they want their film to be successful. Bottom line: *Baghdad ER* is a quite reasonable depiction of not only ER life specifically in Baghdad's Green Zone (circa 2005), but of life in a busy ER in general. The nonmedical viewer will come away with a good appreciation of what medical staffs in those arenas face day in and day out.

John Elliott is the pen name of a US government physician who served in Iraq.

BODY OF LIES

V. L. Vorbeck

Body of Lies, based on the novel by *Washington Post* national security correspondent David Igna-tius, is an interesting film in its examination of natural tensions at work in intelligence. Chief among these in this film are relations with liaison services, specifically those who arc closc to us on the "front lines" of counterterrorism. In this movie the US operatives clearly demonstrate two schools of thought on liaison relations. Russell Crowe's character Ed Hoffman, a CIA Headquarters-based officer, exhibits a nearly heartless unilateralism in which he is not shy about throwing American weight around to get what he wants. In discussing access to a key penetration asset that our Jordanian allies are controlling, he crassly throws out the question, "Who pays the bills?" Contrast Hoffman with Leonardo DiCaprio's more sympathetic character, Roger Ferris, who speaks Arabic, shows deference to the counterterrorism chief of the Jordanian service, and (literally) hands him the file on a target of mutual interest. Ferris represents the Lawrence of Arabia–type who identifies with the locals and seeks optimum results by trusting them explicitly. He and Hoffman are in constant conflict as they maneuver to penetrate a particularly virulent al-Qa'ida-type terror group.

- **Release Date:** October 2008
- **Director:** Ridley Scott *(Gladiator, Black Hawk Down)*
- **Film Company:** De Line Pictures; Scott Free Productions; Warner Brothers (US distributor)
- **Screenwriters:** William Monahan (from a David Ignatius novel)
- **Domestic Gross:** $39.4 million
- **Foreign Gross:** $75.9 million

While the director of this film, Ridley Scott, has us leaning heavily toward the Ferris approach, he does not completely discredit Hoffman and his more Machiavellian approach. Hoffman may not be sympathetic to any cultural mores or attempt to create personal friendships that might enhance the mission, but he is effective in his own way. Given Ferris's deep immersion into the local culture, bosses like Hoffman would have valid concerns over control of the operation and the independence of their case officer.

An entertaining, perhaps intentionally tongue-in-cheek, aspect of the film are scenes in which Hoffman conducts his spy direction from his home, car, kid's soccer practice, etc. This is either Hollywood's version of showing virtual control of operations worldwide or part of a recruitment campaign by the CIA to show that the chief of a major division can save the world and still have quality time with family. It also suggests for generation Y and beyond that one can physically knock a boss over a chair, call him a "fat ass" and still expect full (well, maybe partial) support for your operation.

While the film was entertaining, it had its "can you believe he is doing this" moments. For me it was when Ferris, coming off a failed marriage, decides to enter a Palestinian refugee camp in Jordan to romantically pursue a nurse (of Iranian descent). For a man on the verge of penetrating a deadly terror group this seems beyond dangerous. Ferris moves about the camp with ease, relying on his Arabic and American naiveté, to sustain him in an area teeming with Hamas, PLO, and other operatives as well as a host of foreign intelligence services. Given the physical safety and CI implications it is more than surprising he is not really taken to task for it by the Jordanians or Americans. Such is Hollywood.

V. L. Vorbeck is the pen name of a military veteran with six years work in CIA's Counterterrorism Center.

The Recruit

John Anderson, Clifford L. and Lucy B.

Although fun to watch, not much of what happens in *The Recruit* actually dovetails with my experience in training to become a certified case officer. The training sequences in the movie focus on disassembling weapons, hand-to-hand combat, and breaking into houses—the kind of stuff we rarely, if ever, really do. Very little time in this portrayal was devoted to the basics of tradecraft that in fact make up much of what we do: time-consuming SDRs (surveillance detection routes), casual elicitation, and report writing. That being said, a movie focusing on basic tradecraft would not be much of a box office draw because these basic work activities can in fact be quite pedestrian, so I can understand why the movie was made the way it was.

- **Release Date:** January 2003
- **Director:** Roger Donaldson *(Cocktail, Thirteen Days)*
- **Film Company:** Touchstone Pictures; Spyglass Entertainment; Epsilon Motion Pictures; Buena Vista Pictures (US distribution)
- **Screenwriters:** Roger Towne (*In the Company of Spies*); Kurt Wimmer
- **Domestic Gross:** $52.8 million
- **Foreign Gross:** $48.4 million

> Personally, my main motivation in joining the Agency was to work against the terrorist target. For this reason, *The Recruit* did not really influence my decision to apply—very little of the movie dealt with issues that were important to me. I did find the fast-paced lifestyle portrayed in the movie appealing, however. I would recommend the movie to an aspiring NCS candidate, but only for its entertainment value.
>
> —John Anderson

The Recruit is an entertaining movie. I watched it before joining the Agency. I found it interesting then and wondered how much of it was true and how much was Hollywood. I particularly liked the DVD special features that included minidocumentaries about the CIA, hosted by Chase Brandon, a former operations officer who consulted when the movie was being made. Upon entering the Agency and completing training and certification as a case officer, I still find it interesting, but more as entertainment than as reality. The funny thing is that everyone in the Agency believes the movie is ridiculous but, despite that sentiment, all of the covert service trainees watched the film on the bus going into training and then again back to Washington after graduation.

During every step of our training, many of my fellow trainees would watch this film for "comic relief" from the intense pressures we were experiencing. We would all share a laugh about how melodramatic the movie seemed compared to reality. We scoffed at many inaccuracies regarding the portrayal of our training, for example, when the main characters are discussing whether they would be able to field-strip an assault rifle, blindfolded, during a night jump (that unfortunately was not part of our training). Another favorite laughable inaccuracy was when the NOC (whom they referred to as the NOC, then they would spell out "N-O-C" so the the audience would presumably understand they were referring to nonofficial cover) was assigned as an overt employee at CIA headquarters. Very covert for a nonofficial employee!

> In retrospect, I would suggest other movies or books to an aspiring candidate. One film I would recommend would be *Spy Game*. Often junior officers trying to explain something to another junior officer could be heard saying: "It's like that scene in *Spy Game*, when Brad Pitt…." On the serious side, I would recommend any book by David Ignatius. He writes entertaining spy fiction that hits pretty close to home. I would only recommend *The Recruit* to someone already in the training pool, and even then I would suggest that it should only be viewed when good chuckles are needed (or on the bus to/from training).
>
> —Clifford L.

I watched *The Recruit* when it first came out in about 2003 and again after I completed case officer training more recently. Both times I found the movie entertaining but unrealistic and artificial or "hokey" when compared to real life in the business of espionage. Things that struck me the second time around were the amount of paranoia and questioning (appropriate I now think) that James experiences throughout the movie about what exactly was or wasn't a "test" designed for his training and about his relationship with his instructor/role player. In addition, I found the plot lines less confusing than they seemed when I was a more naive viewer. My second viewing led to several "aha" moments, and specific lines in the movie stood out to me because it was clear that the movie's makers had consultations with CIA officers about what the training entails. They got at least some elements of it spot on, embedded in a lot of Hollywood embelishments.

> To unknowing viewers, this might be an appealing, intriguing, and exciting, if somewhat confusing, movie. Those of us in the business are likely to find some if it accurate, but most of it unrealistic. Now, if only all of us looked like James and Layla; oh, wait, maybe we do and it is just our cover the makes us look like such ordinary people.
>
> —Lucy B.

The authors, all writing in pen name, completed training in 2008.

The Bourne Identity

J. M. Webb

Doug Liman's 2002 film adaptation of Robert Lud-lum's novel, *The Bourne Identity* (1980), is a great action flick that draws the audience in from the get-go and sustains interest with a fast-paced story of intrigue. Fished out of the Mediterranean Sea by strangers, Jason Bourne (Matt Damon) wakes up with no memory of who he is or how he ended up unconscious, riddled with bullets. This sets him on a relentless and dangerous quest of self-discovery. He has only one clue, a numbered account in Zurich found on a subdermal microfilm extracted from his hip by one of those who saved him. The number leads him to a lockbox full of money, a gun, and multiple passports with his photo—but in different names and from different governments. As his quest engages him in one difficult encounter after another, Bourne is as spellbound as the audience as he inexplicably and successfully engages in precise hand-to-hand combat with local police and Marine guards trying to detain him at a US embassy.

- **Release Date:** June 2002
- **Director:** Doug Liman (Mr. and Mrs. Smith)
- **Film Company:** Universal Pictures; Hypnotic; Kalima Productions GmbH & Co. (German); Kennedy/Marshall Company; Stillking Films
- **Screenwriters:** Tony Gilroy and W. Blake Herron
- **US Domestic Gross:** $121.7 million
- **Foreign Gross:** $92.4 million

Not knowing where his paramilitary skills came from makes Bourne uneasy and wary of himself, but he nevertheless trusts his tradecraft—and a girlfriend he picks up along the way—which helps him survive potentially fatal encounters. He is so good that he isn't "getting off the X" (an ambush or attack point) so much as he is seeing it well in advance and turning it to his advantage. As he accumulates clues to his identity, Bourne becomes disgusted and horrified at the prospect that he is an assassin for the Central Intelligence Agency. When he realizes the truth, all he wants is out. Others want him dead.

The "others" are less-than-upstanding officers in a CIA with extraordinary powers, who have concluded that Bourne is a "malfunctioning, $30 million black ops catastrophe." Lurking in dark rooms and operating well outside ethical and legal norms, these officials are transfixed with taking Bourne out for having failed to discreetly kill a deposed African leader who threatened to "name names" in a tell-all book about the CIA's activities in Africa if the CIA didn't put him in back in power. By the end of the movie Bourne has slipped through the Agency's fingers and a senior Agency official is left trying to tie off the problem by killing subordinates who know about the program.

Liman's screenplay of *The Bourne Identity* stays fairly close to Ludlum's original storyline and makes for a great time with good reason. The first of the Bourne conspiracy books—which also include *The Bourne Supremacy* (1986) and *The Bourne Ultimatum* (1990)—it was voted the second best spy novel by *Publishers Weekly,* following John le Carré's *The Spy Who Came in from the Cold.* Bourne's jagged discovery of who he is and the dawning of emotional conflict with his training make this spy thriller timeless. Tight shots on the fight and chase scenes in the film adaptation don't hurt either, but the mystery of who Bourne is and his struggle with himself keep viewers engaged.

The movie's strong undertone that political assassination is reprehensible gives one something to ponder. US government employees have been prohibited from conducting assassination on behalf of the government since the Ford administration issued Executive Order 11905 in February 1976. Two years later, the Carter administration expanded that order to encompass any persons acting on behalf of the US government, a position the Reagan administration reaffirmed not long after *The Bourne Identity* appeared in 1980. The issue is less straightforward in the wake of al-Qa'ida's declaration of war on the United States and our effort to fight terrorism, raising the question "do attempts to kill terrorists leaders in their lairs constitute assassination or are they justifiable acts of war?" Ethical questions—individual, organizational, and national—undoubtedly arise as CIA officers are asked to assume personas, bend the laws of other states, or take quiet actions on behalf of the President in pursuit of national security. Part of what makes watching *The Bourne Identity* interesting is seeing characters so deeply flawed they think they can act with impunity in the pursuit of their own self interests, particularly since the Agency works so hard to give officers a strong ethical compass to guide them through murky waters. What's disquieting however, is when CIA officers are caught in the middle of the pursuit of national interests, debates over the national conscience, and clashing political philosophies.

J. M. Webb is the pen name of an analyst in CIA currently assigned abroad.

BURN NOTICE

Lisa M. Forrester

Burn Notice is a highly entertaining, action-filled television series about a former spy, Michael, who was "burned" or discredited by someone (not further identified) in his former organization and basically hung out to dry. He spends his time doing unusual jobs for hire, fighting criminals and getting people out of trouble, while all the time trying to find out who burned him and why. He's got two sidekicks, a gorgeous ex-girlfriend and a former colleague, who feature in each episode. They are all highly trained operatives who can build explosives in their kitchens; handle most any kind of vehicle, weapon, or piece of technology with aplomb; and play any role that is required of them in their efforts to complete their "jobs." The show, filmed in Miami, is glitzy, glamorous, and portrays the role of a spy as exciting, spontaneous, and mostly successful—the spies generally manage to get the bad guys killed and the mostly innocent victims out of trouble.

- **Release Date:** May 2007 (USA Network)
- **Director:** Various
- **Film Company:** Fox Television Studios; Fuse Entertainment
- **Screenwriters:** Various
- **Domestic Gross:** NA
- **Foreign Gross:** NA

Michael has built a network of contacts in all areas of the Miami underworld and gets what he needs in the way of weapons, money, and supplies. The value to an operative of such a network and the skills needed to acquire it are highlighted. The show's writers illustrate the value of good information, intelligence, in any operation, showing the detailed files Michael compiles on his targets and the types of data he collects—their habits, behaviors, relationships, overseas contacts, funding, cover stories, and backgrounds. He is highly skilled at developing and maintaining professional relationships and has the complete trust of his two partners.

Burn Notice does a credible job of portraying the less glamorous aspects of espionage work. For example, Michael's family lives in Miami and frequently involves itself in his life, but he is torn between his desire to maintain a relationship with them and his need to avoid getting too close to keep them from becoming embroiled in his activities and to avoid opening himself up to them. The show thus offers a glimpse of the isolation an operative can feel when he has few confidants and is used to working independently; this inability to form solid personal relationships is one of the main plot themes.

At the same time, the show can be humorous, as the clever writers continually come up with apparently impossible situations from which Michael and his partners must extricate themselves. Plan A rarely ever works—thus showing the need for backup plans. *Burn Notice* also underlines the importance of help: Michael rarely ever succeeds without the help of his two friends. In demonstrating the need in espionage for such attributes as flexibility, adaptability, and ability to think on one's feet, *Burn Notice*—even while largely written tongue-in-cheek—depicts spies in a positive light. On top of that, good prevails over evil in nearly every episode, a welcome portrayal of espionage for a change!

Lisa M. Forrester is a pen name for a CIA case officer with 23 years of service who is currently stationed in Asia.

TAKEN

Shirley A. Healer

In this highly entertaining—on many levels quite implausible—unexpectedly touching movie, Liam Neeson plays a retired government covert agent, Bryan Mills, whose ambitions as a CIA operative led to the demise of his marriage and to a distant relationship with his teenage daughter, Kim (played by Maggie Grace). In an attempt to develop a closer and more trusting relationship with Kim, Mills reluctantly and against his better judgment allows her to travel to Paris with one of her teenage friends.

Shortly after their arrival, Kim and her friend are abducted by an Albanian drug trafficking organization known for prostituting captured girls. Of course it follows that Mills must reengage in operational work to rescue her. The hunt for the abductors is action-packed and keeps the adrenaline flowing. The movie is filled with the stunts, gadgets, and spy "tools" that we all love to fantasize about. However, just as you will be sitting on the edge of your seat waiting for the next chase, you will find the movie suddenly touching because Neeson also does an excellent job of portraying the more vulnerable paternal side of his character. The vacillation between watching Nee-son employ exciting tradecraft techniques and experiencing his despair and loneliness as he frantically searches for his daughter makes the film entertaining, exhilarating, and at the same time emotional.

- **Release Date:** January 2009
- **Dircctor:** Picrre Morel
- **Film Company:** Europa-Corp; M6 Films; Grive Productions; All Pictures Media; Papillon Productions; Wintergreen Productions
- **Screenwriters:** Luc Besson and Robert Kamen
- **Domestic Gross:** $144.6 million
- **Foreign Gross:** $76.2 million

Although based on unrealistic operational and relationship scenarios (most of us are not granted the opportunity to fix our damaged family relationships by becoming a one-time hero), the movie touches upon the hidden and quite common fantasy that a person can remedy significant family upheavals resulting from challenging and time-consuming careers or other distractions by ironically employing unique work-related knowledge and skills to rescue the estranged loved ones. Furthermore, Hollywood's propensity to glorify the world of espionage as dangerously sexy—but hidden from most people living mundane lives—appeals to our fantasies on another level. While at times the world of espionage is indeed both dangerous and exciting, our loved ones tend to find our opaque and mysterious work unpredictable and frustrating—even if in real life our work and the skills we employ are often not all that far removed from those employed by ordinary people in ordinary, fully transparent jobs. In this sense the price Neeson's character pays in his personal life refers to very real dangers to relationships caused by a covert lifestyle, not to wildly overimagined operational dangers portrayed in the movie.

In the end, therefore, *Taken* indulges us with fantastic tradecraft, dangerous psychopathic villains, and epic operations and portrayals of awed and grateful family members that have very little connection to reality, yet are very fun to imagine. But that is what makes this a Hollywood movie.

Shirley A. Healer is the pen name of a CIA psychologist deployed to CIA's Counterterrorism Center.

INDEX

A

B

C

S

T

U

V

W

X